故園畫憶

庚寅中秋
韓磐陀題

《故园画忆系列》编委会

名誉主任： 韩启德

主　　任： 邵　鸿

委　　员：（按姓氏笔画为序）

万　捷	王秋桂	方李莉	叶培贵
刘魁立	况　晗	严绍璗	吴为山
范贻光	范　芳	孟　白	邵　鸿
岳庆平	郑培凯	唐晓峰	曹兵武

故园画忆系列
Memory of the Old
Home in Sketches

黑土地上的俄式风情
Russian Style on the Black Land

王安安　绘画 撰文
Sketches & Notes by Wang Anan

学苑出版社
ACADEMY PRESS

图书在版编目（CIP）数据

黑土地上的俄式风情 / 王安安绘画、撰文. -- 北京：学苑出版社，2013.6

（故园画忆系列）
ISBN 978-7-5077-4287-9

Ⅰ.①黑… Ⅱ.①王… Ⅲ.①建筑艺术 - 钢笔画 - 作品集 - 中国 - 现代 Ⅳ.①TU-881.2

中国版本图书馆CIP数据核字(2013)第105874号

出 版 人：	孟 白
出版发行：	学苑出版社
社　　址：	北京市丰台区南方庄2号院1号楼
邮政编码：	100079
网　　址：	www.book001.com
电子信箱：	xueyuan@public.bta.net.cn
销售电话：	010-67675512、67678944、67601101（邮购）
经　　销：	全国新华书店
印 刷 厂：	三河市灵山红旗印刷厂
开本尺寸：	889×1194　1/24
印　　张：	5
字　　数：	120千字
图　　幅：	122幅
版　　次：	2013年6月北京第1版
印　　次：	2016年1月北京第2次印刷
定　　价：	38.00元

目　录

前　言　　　　　　　　　　王安安

哈尔滨市

圣索菲亚教堂	3
圣索菲亚教堂·建筑布局	4
圣索菲亚教堂·主入口	5
圣索菲亚教堂·侧入口	6
圣索菲亚教堂·帐篷顶与长窗	7
圣索菲亚教堂·立面装饰	8
圣索菲亚教堂·砖质结构	9
圣索菲亚教堂广场	10
圣索菲亚教堂广场·圣灯	11
圣索菲亚教堂广场·灯柱	12
圣阿列克谢耶夫教堂	13
圣阿列克谢耶夫教堂·砖质结构	14
圣阿列克谢耶夫教堂·精致的细部	15
圣阿列克谢耶夫教堂·幽静的环境	16
南岗教堂	17
南岗教堂·朴素的装饰	18
圣母守护教堂	19
圣母守护教堂·拜占庭风格的入口	20
圣母守护教堂·钟楼与穹窿顶	21
秋林商行	22
哈尔滨旅馆	23
哈尔滨旅馆·入口	24
江畔餐厅	25
江畔餐厅·入口	26
江畔餐厅·特色的建筑装饰	27
亚道古鲁布水上餐厅	28
亚道古鲁布水上餐厅·餐厅的陆上部分	29
亚道古鲁布水上餐厅·俄式装饰细部	30
原坎科泰餐厅	31
原坎科泰餐厅·面江的外廊	32
原坎科泰餐厅·精巧的木结构装饰	33
果戈里街	34
花园街	35
花园街·街边小品	36
中央大街	37
中央大街·建筑风格	38
中央大街·街区环境	39
哈尔滨百年老街酒店	40
米尼阿久尔餐厅	41
马迭尔宾馆	42
马迭尔宾馆·建筑布局	43
马迭尔宾馆·建筑风格与特色	44
华梅西餐厅	45
中央大街117-121号	46
松浦洋行	47

松浦洋行·建筑装饰风格	48
松浦洋行·细腻的建筑细部	49
哈尔滨话剧院	50
梅耶洛维奇宫	51
横滨正金银行哈尔滨分行	52
横滨正金银行哈尔滨分行·主入口	53
哈工大土木系教学楼	54
哈工大土木系教学楼·正面入口	55
铁道俱乐部	56
哈医大四院门诊部	57
中东铁路局医院	58
中苏友谊宫	59
中苏友谊宫·主楼建筑风格	60
斯基德尔斯基故居	61
斯基德尔斯基故居·建筑造型	62
斯基德尔斯基故居·沿街立面	63
太阳岛俄式民居	64
太阳岛俄式民居·规整的外部院落	65
太阳岛俄式民居·通透的院墙	66
太阳岛俄式民居·木刻楞房屋	67
太阳岛俄式民居·独立的建筑小品	68
太阳岛俄式民居·俄罗斯式水井	69
太阳岛俄式民居·"苏联人家"	70
太阳岛俄式民居·门厅	71
太阳岛俄式民居·会客及卧室主立面	72

太阳岛俄式民居·居室一角	73
太阳岛俄式民居·火墙子	74

齐齐哈尔市

圣弥勒尔教堂	77
圣弥勒尔教堂·附属建筑	78
火车站俄式站房	79
火车站俄式站房·主立面入口屋顶	80
火车站候车室	81
火车站候车室·主立面及入口	82
火车站木制天桥	83
火车站木制天桥·一侧入口	84
火车站木制天桥·内部构造	85
铁路俱乐部	86
铁路俱乐部·立面构造与装饰	87
昂昂溪俄式住宅	88
昂昂溪俄式住宅·建筑特色	89

海林市

横道河子镇镇区	93
横道河子火车站	94
圣母进堂教堂	95
圣母进堂教堂·环境地势	96
圣母进堂教堂·主厅立面	97
中东铁路机车库	98

中东铁路机车库·车库门	99	俄式木屋	103
中东铁路机车库· 　破败的建筑与环境	100	俄式木屋·木质结构和屋顶	104
		民居院落·栅栏院	105
中东铁路机车库·区域入口道路	101	民居院落·粗制的院门	106
中东铁路大白楼	102	民居院落·讲究的外环境	107

Contents

Author's Preface — Wang Anan

Harbin City

Saint Sophia Church	3
Saint Sophia Church · Layout	4
Saint Sophia Church · Main Entrance	5
Saint Sophia Church · Side Entrance	6
Saint Sophia Church · Tent Roofs and Long Windows	7
Saint Sophia Church · Facade Decoration	8
Saint Sophia Church · Brick Structure	9
Saint Sophia Church Square	10
Saint Sophia Church Square · Holy Lamp	11
Saint Sophia Church Square · Lamppost	12
St. Alekseev Church	13
St. Alekseev Church · Brick Structure	14
St. Alekseev Church · Exquisite Details	15
St. Alekseev Church · Secluded Environment	16
Protestant Church of Nangang District	17
Protestant Church of Nangang District · Simple Decoration	18
Intercession Church	19
Intercession Church · Byzantine Style Entrance	20
Intercession Church · Bell Tower and Dome Roof	21
Tyurin Company	22
Harbin Hotel	23
Harbin Hotel · Entrance	24
Riverside Restaurant	25
Riverside Restauran · Entrance	26
Riverside Restaurant · Unque Architectural Decoration	27
Yacht Club, a Waterfront Restaurant	28
Yacht Club · Land Part of the Restaurant	29
Yacht Club · Russian Decorative Details	30
Kanke Tai Restaurant	31
Kanke Tai Restaurant · Verandah Facing the River	32
Kanke Tai Restaurant · Delicately Decorated Wooden Structure	33
Gogol Street	34
Garden Street	35
Garden Street Sketch	36
Central Street	37
Central Street · Construction Style	38
Central Street · Block Environment	39
100-Year-Old Street Hotel of Harbin	40
Miniature Restaurant	41
Harbin Modern Hotel	42
Harbin Modern Hotel · Building Layout	43
Harbin Modern Hotel · Architectural Style and Characteristics	44
Huamei Western Restaurant	45

No.117-121, Central Street	46
Matsuura Company	47
Matsuura Company · Style of Architectural Decoration	48
Matsuura Company · Delicate Architectural Details	49
Harbin Theatre	50
Meerovich House	51
Yokohama Specie Bank, Harbin Branch	52
Yokohama Specie Bank, Harbin Branch · Main Entrance	53
Teaching Building, Civil Engineering Department, Harbin Institute of Technology	54
Teaching building of Civil Engineering Department of Harbin Institute of Technology · Front Entrance	55
Railway Club	56
Outpatient Department, 4th Affiliated Hospital of Harbin Medical University	57
Affiliated Hospital of Middle East Railway Bureau	58
Sino-Soviet Friendship Palace	59
Sino-Soviet Friendship Palace · Architectural Style of the Main Building	60
Former Residence of Skidelsky	61
Skidelsky House · Architectural style	62
The Skidelsky House · Street Facade	63
Russian Houses on Sun Island	64
Russian Houses on Sun Island · External Courtyards	65
Russian Houses on Sun Island · Latticed Courtyard Walls	66
Russian Houses on Sun Island · Wooden houses	67
Russian Houses on Sun Island · Architectural Sketches	68
Russian Houses on Sun Island · Russian Style Wells	69
Russian Houses on Sun Island · "Soviet Family"	70
Russian Houses on Sun Island · Foyer	71
Russian Houses on Sun Island · Parlor and Bedroom	72
Russian Houses on Sun Island · Bedroom	73
Russian Houses on Sun Island · Huoqiangzi	74

Qiqihar City

St. Michael's Cathedral	77
St. Michael's Cathedral · Outbuildings	78
Russian Railway Station House	79
Russian Railway Station House · Roof of Main	80
Railway Station Waiting Room	81

Railway Station Waiting Room · Main Facade and Entrance	82	Church of the Presentation of the Blessed Virgin Mary · Main Facade	97
Railway Station Wooden Footbridge	83	Engine Garage, Middle East Railway	98
Railway Station Wooden Footbridge · View from one End	84	Engine Garage, Middle East Railway · Garage Doors	99
Railway Station Wooden Footbridge · Internal Structure	85	Engine Garage, Middle East Railway · Dilapidated Building and Environment	100
Railway Club	86	Engine Garage, Middle East Railway · Road of Regional Entrance	101
Railway Club · Facade and Decoration	87	Dabailou (Grand White Building), Middle East Railway	102
Russian Dwellings in Ang'angxi	88	Russian Wooden Houses	103
Ang'angxi Russian Dwellings · Architectural Features	89	Russian Wooden Houses · Wooden Structure and Roof	104

Hailin City

Downtown Hengdaohezi	93	Residential Courtyards · Fence Courtyard	105
Hengdaohezi Railway Station	94	Residential Courtyards · Rustic Courtyard Gates	106
Church of the Presentation of the Blessed Virgin Mary	95	Residential Courtyards · Exquisite External Environment	107
Church of the Presentation of the Blessed Virgin Mary · Environment	96		

前　　言

　　2009~2010年，由于生活和工作的需要，我远赴中国黑龙江省，对包括哈尔滨在内的3个地区进行了总计长达4个多月的实地考察和调研。特有的自然环境，不同的历史发展，以及与周边国家地区的交流使东北地区的建筑景观具有浓郁的异国色彩。挺拔雄伟的教堂、异域风情浓烈的商场、作为区域视觉重心的火车站，以及色彩明艳、工艺精湛的俄式民居……所有的一切都留给了我难以磨灭的深刻印象。惊艳于它们的美，惊诧于它们的细腻。于是在工作之余，我用手中饱含激情的笔在画纸上记录下关于东北地区建筑景观的视觉信息，最终有了100幅画作。

　　2012年，经学苑出版社周鼎老师的推荐，我有了参与编写《故园画忆》系列书籍的机会。抑制不住心中强烈的念想，毫不犹豫选择了东北各地区建筑景观环境作为表现与介绍的对象，一方面希望能够将特色的建筑文化展现出来，另一方面也借此抒发自己的"忆"之情怀。

　　本书所介绍的3个地区各具特色，却又统一于东北地区大文化氛围中。第一部分：哈尔滨市，分别介绍了代表性宗教建筑、商业建筑、文化街区、观演和服务性建筑、民居等。第二部分：齐齐哈尔市及昂昂溪区，主要介绍当地具有建筑特色的教堂、火车站、住宅等。第三部分：海林市横道河子镇，以镇区内历史建筑为主要研究对象。通过客观叙述和速写表现，能够使读者从一个侧面了解地区历史、民族构成、建筑特征、宗教信仰、人民生活习惯等各方面信息。

　　本书完成之际，特别感谢那些帮助我的人们：学苑出版社的领导，他们为我提供了编写出版的机会；我的学生们，他们给了我很多有益的意见和建议。感谢我的家人、朋友和一切帮助我的人。

王安安

2013年3月

Foreword

From 2009 to 2010, for more than four months, for my work, I went to 3 regions of China's Heilongjiang Province, including Harbin, to do on-the-spot investigation and research. The unique natural environment of the northeast region, the different historical development situation and influences from neighboring countries have created a strong distinctive architectural landscape - such as the tall, majestic churches, exotic shopping malls, a train station which is regarded as a regional visual center, and colorful Russian houses with excellent construction aspects. All these impressed me unforgettably. Amazed by their beauty and delicacy, I used my pen to passionately record the visual information of architecture of the northeast region in my spare time. Now I have approximately 100 sketches here.

In 2012, as recommended by Mr. Zhou Ding of Academy Press, I agreed to compile a book titled Memory of the Old Home in Sketches. With an uncontrollably strong feeling, I chose, without hesitation, to represent and introduce the architectural landscape of the Northeast. On the one hand, I want to show the cultural characteristics of the architecture. On the other hand, I want to express my memory of them.

Three regions presented in this book are distinctive, but also integrated into Northeast regional culture. In Part I, about Harbin City, religious architecture, commercial buildings, cultural blocks, Performance and service buildings and dwellings are both presented. In Part II, about Qiqihar City and Ang'angxi District, mainly local architecturally distinctive churches, train stations and dwellings are introduced. In Part III, about Hengdaohezi township of Hailin City, historic buildings of the town are the main research object. Objective descriptions and sketches enable readers to understand the history, ethnic aspects, architectural features, religion, and living habits of people in the region.

At the completion of this book, special thanks go to those who helped me: leaders of Academy Press who provided me the opportunity to compile and publish; my students who gave me a lot of useful comments and suggestions. Also thanks to my family, friends and all others who helped me.

Wang Anan
March, 2013

哈尔滨市

Harbin City

圣索菲亚教堂

　　位于哈尔滨市道里区，是我国现存最大的东正教堂，俄罗斯古典拜占庭式建筑的典型代表，始建于1923年，1932年建成。全国重点文物保护单位。

Saint Sophia Church

Located in Daoli District, Harbin City, it is China's largest Russian Orthodox Church and a typical example of Russian classical Byzantine style. Construction began in 1923 and was completed in 1932. Now it's a national key cultural relic listed under state protection.

圣索菲亚教堂·建筑布局

平面呈东西走向的十字架形状，有四个门出入，正门顶部为钟楼。屋顶为一个巨大饱满的洋葱头穹顶统率着四翼大小不同的帐篷顶，形成了主次分明的布局。

Saint Sophia Church • Layout

The plane is east-west cross shape with four doors. At the top of the main entrance is a bell tower. The roof is a huge, onion shaped dome commanding the four wings of different sizes of tent-shaped tops.

圣索菲亚教堂·主入口

主入口部分突出于教堂的主体结构，以不同的组织方式构成的同心圆弧型层叠错落，两侧对称出现的红砖组成的十字型，象征着与天堂诸神心意相通的精神气质。

Saint Sophia Church • Main Entrance

It stands out with varied concentric circle arcs. Bilaterally symmetrical red bricks form crosses to symbolize the connected spirit with the mind of God.

圣索菲亚教堂·侧入口

侧入口作为一个嵌入型构成元素与其周边的多层线脚完美融合，共同为组成多面体的建筑结构服务。

Saint Sophia Church • Side Entrance

It is perfectly integrated with surrounding multilayer moldings as an embedded constituent and contributes to form a polyhedron structure.

| 圣索菲亚教堂·帐篷顶与长窗 |

帐篷顶出现在建筑的四翼，既是洋葱头穹顶的辅助图形，又在一定程度上呼应了圆柱形的主体结构，墨绿色的铁皮覆面丰富了整体建筑的色彩层次。

Saint Sophia Church·Tent Roofs and Long Windows

The tent roofs on the four wings of building support graphics of the onion dome and echo the cylindrical structure to a certain extent. Clad in dark green metal, the roofs enrich the color of the entire building.

圣索菲亚教堂·立面装饰

　　侧立面分为三段：上段是红砖造型层次最为丰富的部分；中段由套叠凹入的矩形包裹连续排列的十字形组成；下段厚重稳固，显示粗犷之气，与建筑主体的精雕细刻形成了强烈的对比。

Saint Sophia Church · Facade Decoration

The side facade is divided into three sections. The upper section is the part of a rich red brick molding. The middle consists of continuous crosses arranged in rectangular recesses. The lower part is thick and plain and forms a strong contrast to the exquisite details of the structure.

圣索菲亚教堂·砖质结构

  教堂红砖结构从图形关系上来看，所用的艺术处理方法有重复、对比、韵律、比例等，最终形成的复杂结构统一于整体建筑的框架之中，可谓是"精之有道，砌之有体"。

Saint Sophia Church • Brick Structure

In terms of the graphical aspects of the red brick church structure, the artistic options used here include repetition, contrast, rhythm and proportion which ultimately combine to integrate the complex structure a perfect whole. It can be described as "the proper way to build in a perfect structure".

【圣索菲亚教堂广场】

　　教堂广场整体风格别具欧式特征，在广场中视线几乎毫无遮挡，平缓的地面抬起向中心处升高，周围围合有各具特色的商业建筑、历史建筑等。

Saint Sophia Church Square
The overall square is in the unique European style with an almost unobstructed view. The flat ground rises toward center and is surrounded by distinctive commercial and historic buildings.

圣索菲亚教堂广场·圣灯

广场内部的圣灯代表了一种宗教情结，和教堂相应，体现着这片景观环境特有的精神气质。

Saint Sophia Church Square • Holy Lamp

The Holy Lamp in the square represents a religious complex and match the church very well, embodying the ethos of this unique landscape environment.

圣索菲亚教堂广场·灯柱

灯柱是广场上代表性元素，黑色的铸铁灯托上浮现出具有象征意义的花卉和标志浮雕；层次丰富的线脚，还有四面体现历史中英雄主义人物形象的浮雕画。

Saint Sophia Church Square· Lamppost

The lamppost is a representative element of the square. Symbolic flowers and flags emerge from the black cast iron lamp bracket; the moldings in rich layers offset four relief paintings of heroic historical characters.

圣阿列克谢耶夫教堂

　　又称圣母无染原罪堂，1935年落成，建筑面积1005平方米，占地面积3000平方米，是一座保存完好的俄罗斯风格的拜占庭式教堂。1980年起重新开放。

St. Alekseev Church

Also known as the Church of the Immaculate Conception, completed in 1935 with a construction footprint of 1,005 square meters in an area of 3,000 square meters. It is a well-preserved Russian Byzantine style church re-opened to public in 1980.

圣阿列克谢耶夫教堂·砖质结构

教堂由新旧两部分共同组成：一部分木结构建筑，一部分砖石结构建筑。外观别致新颖，墙面线条丰富。

St. Alekseev Church · Brick Structure
The church is composed of old and new parts - one wooden, the other masonry. It has a novel appearance with rich wall.

{圣阿列克谢耶夫教堂·精致的细部}

  教堂拥有耐心寻味的精致细部：阴阳交错的线脚与连拱装饰共同营造出明暗关系明确的檐部效果；建筑转角处设置有韵律出现的阴刻凹槽。

St. Alekseev Church • Exquisite Details

The church has tasteful exquisite details.
The staggered architraves and arch decorations create eaves with clear light and shade effect with carved intaglio.

圣阿列克谢耶夫教堂·
幽静的环境

教堂周围用绿地围合，绿地边缘用曲线花岗石板做成休闲椅，给人以宁静的感觉。教堂外墙体整修粉饰后，闪着历史的"光芒"。

St. Alekseev Church • Secluded Environment

The church is surrounded by belts of greenery the edges of which Are marked with chairs of curved granite panels, all creating a sense of tranquility. After renovation, the external walls of the church shine with the "light" of history.

{ 南岗教堂 }

　　位于南岗区东大直街252号，于1916年建成，砖木结构，两层楼房，平面呈拉丁十字形。临街设置主入口，上方为钟楼，为仿中世纪哥特式建筑风格。

Protestant Church of Nangang District

It's at 252 Dongdazhi Street, Nangang District.

This two-story brick and wood building was built in 1916.The plane is in Latin cross shape and the main entrance faces the street. There is a bell tower on top in the medieval Gothic architectural style.

尼埃拉依教堂·朴素的装饰

　　教堂外形朴实无华，外壁极少雕饰物，突出了哥特式的尖拱屋顶。为了减轻屋顶的重量采用铁皮盖，加厚了墙壁，减少了层叠的窗户，以此适应了哈尔滨气候寒冷的特点。

Protestant Church of Nangang District • Simple Decoration

The appearance of Church is plain with very few carvings on outside walls thus highlighting the Gothic arch roof. In order to reduce the weight of the roof, a metal cover was used; walls were thickened and the cascading windows are smaller to adapt to the cold Harbin climate.

〔圣母守护教堂〕

又称圣母帡幪教堂，1922年动工建造，木质结构。1930年在原基础上重新建造了现在的砖石结构教堂，是仿照土耳其伊斯坦布尔的圣索菲亚大教堂的艺术风格而建造的。

Intercession Church

Also known as Ping Meng Church of Notre Dame. It was built in 1922 as a wooden church. Present day masonry church imitating the style of Hagia Sophia in Istanbul, Turkey was in 1930 in place of the original one.

圣母守护教堂·拜占庭风格的入口

教堂平面呈希腊十字形，体现出拜占庭建筑的特点。建筑入口处踏步采用白色石基，和教堂整体红砖形成强烈对比，上方门头重复出现3个正十字形，体现建筑的宗教功能。

Intercession Church • Byzantine Style Entrance

The shape of the Church is a Greek cross. It reflects the characteristics of Byzantine architecture. White stone is used for the steps of the main entrance and is in sharp contrast with the red bricks of the church. Above the door, are three crosses reflecting religious function of the building.

圣母守护教堂·
钟楼与穹窿顶

　　中央大穹窿顶控制着整个教堂结构，下部用帆拱承托上部圆顶。大穹顶鼓座有12洞花窗，入口处上方的钟楼又起一处小穹顶，错落有致，凝重庄严。

Intercession Church • Bell Tower and Dome Roof

The central dome roof dominates the entire tructure. There are 12 flowered windows around the circumference of the dome. There is also a small dome roof atop the entrance bell tower. The structure of well arranged and dignified.

[秋林商行]

  旧时俄商在哈尔滨开办的最早的商行之一,建筑以深沉的暗绿色调为主色,有着优雅的装饰细部和精美的穹顶造型,平面布局简洁。

Tyurin Company

One of the oldest Russian firms in Harbin, it is dark green as and has elegant decorative details and an exquisite dome. The layout is simple.

哈尔滨旅馆

　　现为哈尔滨国际饭店，始建于1936年，由俄国著名建筑设计师设计，其外型设计为"手风琴"式建筑，是新艺术运动建筑风格在哈尔滨的代表作。

Harbin Hotel
Known as Harbin International Hotel, it was designed by a famous Russian architect and built in 1936. Its exterior design is accordion shape and is a masterpiece of Art Nouveau architecture in Harbin.

哈尔滨旅馆·入口

　　建筑具有典型的新艺术运动建筑风格特征：在建筑设计上追求简单纵横直线的形式倾向，采用简单的立体式几何形式，加上形如植物卷须般平滑的自然曲线进行装饰。

Harbin Hotel · Entrance

The building has typical Art Nouveau architectural characteristics using simple vertical and horizontal straight lines. using simple three-dimensional geometric forms and smooth natural curves shaped like plant tendrils for decoration.

江畔餐厅

位于道里区斯大林公园内，原为铁路松花江站站舍之一，建于1930年，单层木质结构，俄罗斯风格建筑。

Riverside Restaurant

Located in Stalin Park, Daoli District. It was formerly one of the Songhua River Railway Station dormitories, It is a single storey Russian style wooden structure built in 1930.

| 江畔餐厅·入口 |

建筑入口处设高大台阶,有装饰性木柱、栏杆和山墙板。木门窗套顶端采用曲线的装饰纹样,建筑的檐口做多层锯齿状花纹,柱身制作十分复杂。

Riverside Restaurant • Entrance

There are high steps located at the entrance of the building with decorative wooden pillars, railings and gable boards. On top of the window cover of wooden door, are curved decorative patterns. The cornice of the building is in multi-layered jagged pattern with an extremely complex pillar.

江畔餐厅·特色的建筑装饰

建筑的北立面（沿江立面）为主立面，中轴对称，屋顶成十字形，在哈尔滨近代建筑中独一无二。柱间为木板栏杆，檐部雕花，栏板与柱的设计浑然一体。

Riverside Restaurant • Unique Architectural Decoration

The northern facade of the building (along the river) is the main façade with axial symmetry and cross-shaped roof. It is unique in Harbin Modern Architecture. Between the columns are wood railings. The carved eave integrates seamlessly with the columns.

[亚道古鲁布水上餐厅]

位于道里区九站街比邻的松花江畔，建于1912年，占地面积4550平方米，设计师为米扬高夫斯基，木结构，俄罗斯建筑风格。

Yacht Club, a Waterfront Restaurant

Located beside Songhua River which is next to Jiuzhan Street, Daoli District, it was built in 1912, covering an area of 4,550 square meters. The traditional Russian wooden style building was designed by M. Yankovsky.

亚道古鲁布水上餐厅·餐厅的陆上部分

哈尔滨的标志性建筑之一。有着独树一帜的建筑艺术，在世界上是独一无二的。它依堤傍水，舒展开敞，一部分建于江堤上，一部分深入江中。

Yacht Club·Land Part of the Restaurant

It is unique for its unique architecture design and is one of the famous landmarks of Harbin. Built on the embankment, it extends out over the river.

亚道古鲁布水上餐厅·俄式装饰细部

建筑外围院落大门及附属用房均有着典型俄式装饰细部：几何式外框；艳丽的色彩搭配；不同装饰形式的出檐木线脚；以竖向木板条为外部装饰的建筑外墙体。

Yacht Club · Russian Decorative Details

The courtyard gate and the rooms have typical Russian decorative details including geometric frame, bright colors, eaves and architrave in different decorative forms and the external wall of the building decorated by vertical slats.

原快科泰餐厅

坐落在斯大林公园内的松花江畔,砖木结构建筑。建于1930年,设计师是日本人,建筑既有浓郁的俄罗斯民间风格,又有日本的建筑风情。

Kanke Tai Restaurant

It is a wood and brick building located beside the bank of the Songhua River in Stalin Park. It was built in 1930 by a Japanese designer. The building has both rich Russian folk style and Japanese architectural style.

原坎科泰餐厅·面江的外廊

餐厅面江的曲折外廊是人们最常停留的地点,可倚着白色镂空的栏板欣赏江上的风光。上部是人字形双坡屋面,有着由数层不同形式和色彩的装饰组成的精美檐部。

Kanke Tai Restaurant • Verandah Facing the River

The zigzag verandah of the restaurant which faces the river is the favorite place for people to leaning against the white hollow railing to enjoy the river scenery. The upper part has a herringbone double sloping roof with exquisite eaves with several layers of decorations in different forms and colors.

**原坎科泰餐厅·
精巧的木结构装饰**

餐厅最显著的特点体现在木结构的装饰上，孔雀开屏样式的木柱头，与俄罗斯传统的尖坡顶结合得十分巧妙。柱身大面积的白色配合黑色边框，在整体建筑的色彩对比中最为醒目。

**Kanke Tai Restaurant
· Delicately Decorated
Wooden Structure**

The most striking feature of the restaurant is the decoration. The top of the wooden pillars with spread peacock tail combines tactfully with a traditional Russian sharp slope roof. A large area of white on the pillar setting off the black frame is the most striking color contrast of the entire building.

> 果戈里街

　　建于1901年，1902年秋林公司在该街盖起大楼，随后围绕秋林公司左右兴起多家俄国人的商号、药店等建筑。果戈里大街也逐渐成为城市中著名的街道。

Gogol Street

Constructed in 1901, the street got its first structure built by the Tyurin firm in 1902. Soon afterwards, a number of other Russian firms, pharmacies and other buildings were built around the Tyurin Building. Gradually Gogol Street became one of the famous streets of the city.

花园街

　　花园街是哈尔滨市重要的交通干道，它将欧洲新古典主义风格、后现代主义风格、欧洲折中艺术等建筑元素有机结合，充分体现欧式建筑的凝重、大气、浑厚。

Garden Street

Garden Street is an important traffic artery in Harbin. It combines European neo-classical, post-modernist and the European eclectic art architectural styles tactfully and fully reflecting the dignity and majesty of European architecture.

花园街·街边小品

　　花园街的历史文化街区是现存唯一保持哈尔滨开埠时期新城住宅区基本原貌的地区，也是早期俄罗斯风格住宅区的典型代表。沿街的景观小品也显示出厚重的文化积淀。

Garden Street Sketch

The historical and cultural block of Garden Street is the only existing block maintaining the essential original appearance of the residential area during the inception of Harbin. It is also representative of the early Russian style residential areas. The landscapes along the street show the profound accumulation of culture.

中央大街

　　南起经纬街，北至松花江畔的防洪纪念塔广场，全长1450米。街区内现有保护建筑36栋，最早形成于1898年，是伴随着城市发展的一条百年老街。

Central Street

South from Jingwei Street and stretching north to Flood Control Monument Square beside the bank of the Songhua River, it is 1,450 meters long. There are 36 preserved buildings on the street now. It was firstly laid out in 1898 and is a century-old street reflecting urban development.

中央大街·建筑风格

街区内的许多建筑采用了当时流行于欧洲的建筑风格，如文艺复兴、巴洛克、折衷主义、新艺术运动以及俄罗斯式建筑风格，建筑形式异彩纷呈，俨然世界建筑博览会。

Central Street • Construction Style

Many buildings on the street have architectural styles that were popular in Europe at the time, including Renaissance, Baroque, Eclectic, Art Nouveau and Russian. It looks like a World Architecture Exposition of extraordinary splendors of various architectural styles.

[中央大街·街区环境]

　　历经三期城市环境综合整治，中央大街历史文化街区传统风貌得以保存。如今已成为以商业、旅游、休闲、文化、娱乐为主要功能，独具魅力的步行街。

Central Street • Block Environment

After three phases of comprehensive urban environmental improvement, the traditional style of historical and cultural blocks of the central street have been preserved. It has become a charming pedestrian street with functions including commerce, tourism, leisure, culture and entertainment.

哈尔滨百年老街酒店

也称为中央大酒店,原为哈尔滨特别市公署,建造于20世纪二三十年代,折衷主义砖木建筑,共4层,形式组合自由,不讲求固定的法式,比例均衡。

100-Year-Old Street Hotel of Harbin

Also known as the Grand Central Hotel, formerly it was the Harbin Special Municipal Hall. It is a four-storey Eclectic brick and wood building built in the 1920s and 1930s. It is a free form building of balanced proportions of no fixed architectural style.

[米尼阿久尔餐厅]

原为犹太人开办。始建于1926年，1927年落成，砖木结构，新艺术运动建筑风格。共两层，由蓝、绿、白和赭石色组成。为哈尔滨一类保护建筑。

Miniature Restaurant

Originally opened by a Jewish businessman, the brick and wood structure was built between 1926 and 1927. It is a 2-storey Art Nouveau building. It combines blue, green, white and ocher colors and is a first class preserved building of Harbin.

[马迭尔宾馆]

　　位于道里区中央大街89号,原名为马迭尔旅馆,始建于1906年,1913年竣工,具有很高的艺术、科学价值,还是中国近现代许多重要事件的历史见证。

Harbin Modern Hotel

Located at 89 Central Street, Daoli District, formerly known as Ma Diel hotel, was built from 1906 to 1913 with high artistic and scientific value. It witnessed many important events in the history of modern China.

马迭尔宾馆·建筑布局

建筑为砖混结构,地上三层,带有局部阁楼,地下一层,内部装饰豪华典雅,平面功能复杂。建筑的三个沿街立面中,以中央大街立面为主要立面,正中为主入口。

Harbin Modern Hotel • Building Layout

It is a 3 storey building with attic and basement. Though the building is of ordinary function, the interior decoration is luxurious and elegant. The street façade includes three entrances, the middle one being the main entrance.

马迭尔宾馆·建筑风格与特色

建筑属法国新艺术运动建筑风格，造型简洁明快。多姿多彩的女儿墙以砖砌体为主，建筑入口上方或其它局部饰以精巧的阳台，成为了建筑的提神之笔。

Harbin Modern Hotel • Architectural Style and Characteristics

The building is French Art Nouveau style with concise and lively design. The colorful parapet is mainly made of brickwork. Exquisite balconies are decorations above the building entrances are the refreshing aspects of the building.

华梅西餐厅

原名为马尔斯西餐茶食店，始建于1925年，建筑和饮食风格富有浓郁俄罗斯特色，与北京马克西姆西餐厅，上海红房子西餐厅和天津起士林大饭店并称为中国四大西餐厅。

Huamei Western Restaurant

Formerly known as Mars Western Tea Lounge, it was built in 1925. Both its architecture and cuisine are of rich Russian style. It is listed as China's four Western Restaurants together with Maxi Dempsey Restaurant, Shanghai Red House Restaurant and Tianjin Qi Shilin Hotel.

中央大街117-121号

建于1935年，砖混结构，仿文艺复兴建筑风格。墙体采用仿石块砌筑，檐口下设流畅复杂的花饰纹样，转角处设主入口，其上设铁艺栏杆阳台，做工精湛。

No.117-121, Central Street

Built in 1935, the brick structure is Renaissance architectural style. The wall is imitated stone masonry. Smooth complex floral patterns grace the cornices and main entrance. Above the main entrance, is an iron railing balcony of exquisite workmanship.

| 松浦洋行 |

位于中央大街120号，建于1909年，1918年竣工，现在为著名的教育书店，是哈尔滨市最大的巴洛克建筑代表作品，也是中央大街的标志性建筑。

Matsuura Company

Located at No.120, Central Street, it was built between 1909 and 1918. Now it houses a famous education bookstore and is the largest Baroque building of Harbin City. It's also a landmark building of Central Street.

松浦洋行·建筑装饰风格

建筑共4层，外立面华丽生动，装饰复杂，轮廓丰富，通体洋溢着巴洛克奇异生动的效果，体现了巴洛克建筑的动感和力度，几乎囊括了欧洲所有建筑手段的精华。

Matsuura Company· Style of Architectural Decoration

The 4-storey building has an external facade with intricate ornaments and rich outlines. The entire building is filled with the special dramatic effect of the Baroque and reflects the dynamism and strength of Baroque architecture. It includes almost all essences of European construction methods.

松浦洋行·细腻的建筑细部

建筑细部细腻精致，出挑的半圆形花萼状阳台，外凸铸铁曲线栏杆，老虎窗上精致的浮雕，配以优雅的灯饰，彰显出建筑的古朴与典雅。

Matsuura Company · Delicate Architectural Details

The building has exquisite architectural details, an outstanding semi-circular calyx-shaped balcony and convex cast iron railings. The exquisite relief of the dormer windows complement the elegant lighting and highlight of the building's quaintness and elegance.

哈尔滨话剧院

位于道里区兆麟街60号,建于1957年,主体建筑面积为15.5万平方米,共6层,西班牙建筑风格。

Harbin Theatre

Located at 60, Zhao Lin Street, Daoli District, this 6-storey, Spanish style building was built in 1957, with an area of 15,519 square meters.

梅耶洛维奇宫

位于红博广场东南侧，始建于上世纪20年代初，文艺复兴建筑风格，设计者是俄罗斯建筑师日丹诺夫。其第一任主人叫梅耶洛维奇，所以又称作"梅宫"。

Meerovich House

Located at south-east of Hongbo Square, this Renaissance style structure was built in the early 1920's. It was designed by Russian architect Zhdanov; the first owner was Meyer Pavlovic. That's why it was called "Plum (pronounced mei er in Mandarin) Palace."

[横滨正金银行哈尔滨分行]

位于道里区地段街133号，现为黑龙江省美术馆，始建于1912年，1937年建成，建筑为砖混结构，仿古典主义建筑风格。

Yokohama Specie Bank, Harbin Branch

Located at 133 Diduan Street, Daoli District, it houses the Heilongjiang Provincial Art Gallery. It was built in 1912 and completed in 1937. It is a brick structure of the neoclassical style.

横滨正金银行哈尔滨分行·主入口

主入口设在主立面中部两根爱奥尼克式立柱之间。建筑造型严谨，以质朴的块面构成为设计基本手法，通过柱体凹槽和线脚、窗下的多样阴刻纹路表现细部特征。

Yokohama Specie Bank, Harbin Branch· Main Entrance

The main entrance is located between the two Ionic columns in the middle of the main facade. The architectural style is rigorous with basic techniques of using plain block surface and displaying details by cylinder grooves, moldings, and a variety of incised lines under the window.

哈工大土木系教学楼

位于南岗区西大直街66号，现为哈尔滨工业大学建筑学院，1953年建成，设计师为斯维利朵夫。砖混结构，折衷主义建筑风格，哈尔滨市二类保护建筑。

Teaching Building, Civil Engineering Department, Harbin Institute of Technology

Located at 66 Xidazhi Street, Nangang District, it houses the Architecture College of Harbin Institute of Technology. Designed by Sviridov, it was built in 1953. It is brick structure in eclectic architectural style and listed as a Second-Class Preserved Building of Harbin City.

> 哈工大土木系教学楼·
> 正面入口

主入口位于正中，6根仿科林斯柱式的巨型壁柱从三层直通檐下，檐部整齐，层次丰富，配以各种装饰造型形成的层次丰富的阴影，雄伟朴实。

Teaching building of Civil Engineering Department of Harbin Institute of Technology • Front Entrance

The main entrance is located exactly in the middle with six large Corinthian pilasters reaching to the eaves of the 3rd floor. The eave is a variety of rich layers formed by decorative shapes to represent magnificence and simplicity.

铁道俱乐部

　　位于西大直街84号，建于1903年，是仿莫斯科大剧院风格的俄式建筑，哈尔滨市一类保护建筑。室内各种浮雕构思精美、图案典雅。

Railway Club

Located at 84 Xidazhi Street, it was built in 1903 in the Russian style architecture of the Bolshoi Theatre, ingeniously conceived and elegant It is a First-Class Preserved Building of Harbin City .

哈医大四院门诊部

位于南岗区颐园街37号，建筑造型设计汲取欧洲文艺复兴时期建筑风格的特点：通高的立柱、韵律的组窗、夸张的阴影处理方式、醒目的穹顶以及跌落式的处理等。

Outpatient Department, 4th Affiliated Hospital of Harbin Medical University

Located at 37, Yiyuan Street, Nangang District, the building design follows the characteristics of European Renaissance architectural style with tall columns, group windows, exaggerated treatment of shadows, eye-catching dome and drop-treatments.

中东铁路局医院

1907年建，共4层，占地4000平方米，俄罗斯风格建筑。入口立面的连通阳台是建筑的点睛之笔，铁艺护栏造型优雅，线条婉转柔和。

Affiliated Hospital of Middle East Railway Bureau

Built in 1907, this Russian-style, 4-storey building covers an area of 4,000 square meters. The balcony connected to the entrance facade is the final touch to the building. The iron fence is elegant with soft lines.

中苏友谊宫

位于友谊路263号,建于1954年,现改名为哈尔滨友谊宫,是这座城市中为数不多的,中西组合型的大屋顶式建筑之一,由中苏两国的专家通力合作、设计建造。

Sino-Soviet Friendship Palace

Located at 263 Friendship Road, it was built in 1954. It has been renamed the Harbin Friendship Palace. It is one of the few large roof style structures in the city combining Chinese and Western styles . It was designed and constructed by cooperating experts from the Soviet Union and China.

【中苏友谊宫·主楼建筑风格】

主楼正立面体现了俄国古典主义的建筑风格，而两翼的檐头及背部均采用了中国大屋顶琉璃瓦的方式。楼的一面是莫斯科红场的钟楼，另一面是天安门广场上的华表，用以象征中苏友谊。

Sino-Soviet Friendship Palace • Architectural Style of the Main Building

The front facade reflects the classicism style of Russia. The eaves and the back of two wings are Chinese style glazed tile roofs. On one side of the building is a bell tower like that in Moscow's Red Square; on the other side are ornamental columns similar to those in Tiananmen Square. These represent the friendship between China and the Soviet Union. It is one of the landmark buildings of Harbin.

斯基德尔斯基故居

位于南岗区颐园街3号，现为省老干部活动中心，建于1914年，设计师为特罗亚诺夫斯基。砖混结构，地上2层，地下1层，平面对称布局，仿古典主义建筑风格。

Former Residence of Skidelsky

Located at 3 Yiyuan Street, Nangang District, it is currently the provincial Veteran's Center. Designed by Troyanovsky, it was built in 1914. It is a 2-storey brick structure with a basement. It is in the classical architectural style with a symmetrical layout.

斯基德尔斯基故居·建筑造型

　　建筑构图依赖于中轴线，空间组合富有变化，又和谐统一。基本设计元素为直线型体，局部配合曲线型装饰。

Skidelsky House • Architectural style

The architectural composition is as axis with combinations of spaces in harmony and unity. The basic design elements are a linear body with curved decoration.

斯基德尔斯基故居·沿街立面

建筑形式艺术设计手法以韵律为主，窗与壁柱间隔排列。建筑上部凸起，立面中部突出，用大三角形檐与两翼形式分开，拥有弧形檐部的中窗更加强了中轴的控制作用。

The Skidelsky House • Street Facade

The architectural style and art design of this building are rhythmical. Windows and pilasters alternate. The upper part of the building rises above the prominent middle facade. Large triangle eaves separate the two wings; the middle window with a curved eave strengthens the dominance of the central axis.

[太阳岛俄式民居]

　　充满浓郁俄罗斯风情而得以保留的别墅及民宅位于太阳岛风景区的俄罗斯风情小镇内，总27座，是在我国出现较早的俄罗斯格调的别墅群体，形成了独特的自然风貌。

Russian Houses on Sun Island

There are 27 villas and other dwellings in the Russian settlement of Sun Island Scenic Area which preserve the rich Russian style. This Russian style villa appeared earlier in China and has unique natural landscape.

太阳岛俄式民居·规整的外部院落

俄式民居通常都带有独立的外部院落，面积有大有小，与建筑用通透性隔断分隔开，有的种植草皮；有的则稍用室外家具点缀，种植住户喜欢的果树、花草和蔬菜。

Russian Houses on Sun Island • External Courtyards

Russian houses usually have separate external courtyards or varying size. They are separated by latticed partitions. Some courtyards have grass lawns, are paved and ornamented with outdoor furniture and favored fruit trees, flowers and vegetables.

太阳岛俄式民居·通透的院墙

院墙通常为通透式。木栅栏有墨绿色的，也有木板本色的，多在上端锯成锯齿形，中间间隔空隙排列，使院内外的空间得以流通。

Russian Houses on Sun Island • Latticed Courtyard Walls

The walls are usually latticed wooden fences of dark green or natural wood color. This allows for good air circulation in the courtyard.

太阳岛俄式民居·木刻楞房屋

　　镇区内建筑有一部分是木结构的,是用原木搭建的井干式房屋,反映了俄罗斯传统建筑风格,称为"木刻楞"。整体建筑有楞有角,规范整齐,有着冬暖夏凉,结实耐用等优点。

Russian Houses on Sun Island • Wooden houses

Some buildings in the town are well-framed wooden structures made of logs and reflect Russia's traditional architectural style. They are called "wooden houses" or standard shape, solid, durable and cool in the summer and warm in winter.

太阳岛俄式民居·独立的建筑小品

民居外围的庭院经过精心规划与整理，一些生活的附属功能空间也被安排出来，作为小品的形式呈现。例如有些在外部设有单体独立的卫生间或洗澡间。

Russian Houses on Sun Island • Architectural sketches

Some subsidiary function living spaces delicately planned and organized in the outer courtyards are shown in the sketches. For example, separate outside toilets or bathrooms.

太阳岛俄式民居·
俄罗斯式水井

　　一些民居外部院落内设有旧式的俄罗斯式水井，设计考究，结构与工艺基本与与"木刻楞"相一致，反映出了俄民丰富的户外生活。

Russian Houses on Sun Island • Russian Style Wells

Some external courtyards have old-fashioned Russian style wells with the same structure as "Wooden Houses" reflecting the rich, outdoor life of Russians.

太阳岛俄式民居·"苏联人家"

一处以"苏联人家"命名的俄式民居,是1950年俄罗斯百姓生活的缩影。建筑有俄式民居的典型特征,明艳的色彩搭配,还有双层门窗,能够抵御-40℃的严寒。

Russian Houses on Sun Island • "Soviet Family"

A Russian house named "Soviet Family" is a microcosm of the life of the Russians in 1950. It has the typical characteristics of Russian houses: vibrant colors, and double doors and windows enabling those inside to withstand the cold of -40 degrees.

「太阳岛俄式民居·门厅」

　　由建筑入口到内部居室的过渡性空间——门厅通常安排有储物的功能。为了使俄式民居的特色更加突出，会特别放置旧有的生活用品，保留着上世纪人民生活的印记。

Russian Houses on Sun Island • Foyer

A foyer, the transitional entry space between outside and internal rooms, usually also has a storage function. To make the Russian residential features more prominent, old household items are usually placed there to retain the imprint of the people's life in the last century.

太阳岛俄式民居·
会客及卧室主立面

"苏联人家"俄式民居会客及卧室的主立面。古朴的家具让人联想到当时主人平凡的家庭生活。

Russian Houses on Sun Island • Parlor and Bedroom

Parlors and bedrooms of the Russian "Soviet Family" dwellings have. rustic furniture reminiscent of the ordinary family life of the owner.

太阳岛俄式民居·居室一角

"苏联人家"内部居室的一角,在朴素的同时透着浓郁的俄罗斯风情。多数上世纪中期的生活用品被保留下来,成为观赏物。

Russian Houses on Sun Island • Bedroom

The bedroom of a "Soviet Family" dwelling is of rich Russian style. Many of the household items from the middle of the last century are retained as ornamentation.

太阳岛俄式民居·火墙子

中国本土俄式民宅采用了一种特殊的取暖方式：在屋内设置"火墙子"。这是中国地方建筑史上的一大创造。

Russian Houses on Sun Island • Huoqiangzi

China's domestic Russian houses used a special heating: huoqiangzi This was a great creation of local architecture.

齐齐哈尔市
Qiqihar City

{ 圣弥勒尔教堂 }

　　位于龙沙区海山胡同24号，1930年动工，1931年末启用。建筑为钢筋混凝土结构，仿哥特式建筑风格，占地面积1250平方米，主塔通高43米，顶端有高1.75米的十字架。

St. Michael's Cathedral

Located at 24 Haishan Lane, Longsha District, it was built from 1930 to 1931. The building is a reinforced concrete structure of Gothic style. It is 1,250 square meters with a 43-meter-tall main tower topped with a 1.75-meter-tall cross.

圣弥勒尔教堂·附属建筑

　　教堂北侧为总堂主教府，2层黄色楼房，建筑面积360平方米；东北侧有一座2层青砖楼房，亦称小教堂，建筑面积927平方米。

St. Michael's Cathedral • Outbuildings

To the north of the church is the Bishop's House. It is a 2-storey yellow building of 360 square meters. Northeast of the church, is a 2-storey black brick Chapel of 927 square meters.

火车站俄式站房

　　始建于1903年，中东铁路全线通车时建成并正式营业。站房为钢筋混凝土结构，局部木结构，内设有一、二、三等旅客候车室，小卖店、行包房等。

Russian Railway Station House

It was officially opened in 1903 at the same time as the Middle East Railway commenced service. The station house is a reinforced concrete structure with local wood interiors. It is equipped with first, second and third class passenger waiting rooms, small snack shops and luggage rooms.

【火车站俄式站房·主立面入口屋顶】

　　面向站台的主立面最具俄式建筑特征：入口屋顶呈双坡人字形，门廊顶部做单坡，由红色铁皮屋面覆盖，檐部雕花，细部精美绝伦。

Russian Railway Station House • Roof of Main Entrance

The main facade facing the platform has the most Russian architectural features. The roof of the entrance is a double-slope herringbone structure; the top of the porch is in single-slope roof covered with red tin. There are exquisitely detailed carvings on the eaves.

火车站候车室

2004年经重新修缮作为国际旅客候车室，共2层，俄式建筑风格。建筑体块组合明确，秩序井然，材质朴素，色彩简练大气，是不可多得的俄式公共建筑典范。

Railway Station Waiting Room

It became the international passenger waiting room after renovation in 2004. It is a 2-storey Russian style block building. It was built of simple materials and color in a grand style. It is a rare example of Russian public buildings.

火车站候车室·主立面及入口

建筑主立面中部凸出，顶部高起，入口大门上方有弧形雨篷，以红色铁皮覆面，檐部有简单的木质装饰，体现了建筑高超的建造与装饰工艺。

Railway Station Waiting Room · Main Facade and Entrance

The middle of the building's main facade protrudes outward and upward. It has a curved awning covered by red. The eave is decorated in simple wooden pattern. The building reflects a superb construction and decoration process.

火车站木制天桥

滨州线上仅存的木制天桥之一。采用俄式建筑中典型的双坡顶面,以铁皮覆盖,完全依附于桥体抬起与平展的变化。

Railway Station Wooden Footbridge

It is one of a number wooden footbridges along the Binzhou Railway line and is listed as a First Class Preserved Structure of the city. Its typical Russian style double pitch roof is covered with iron sheeting.

火车站木制天桥·一侧入口

　　天桥主要材质以木为主，辅以钢制骨架和铁皮顶面。建筑色彩全部为明度有细微变化的中性灰色，在阳光下形成了完美的素描图底关系。

Railway Station Wooden Footbridge · View from one End

The wooden footbridge is supported by a steel framework and had a metal top. It is a neutral gray color which has subtle changes in the sunlight brightness.

火车站木制天桥·内部构造

立面上部开有方形高窗，使天桥内部拥有良好的自然采光。板、侧壁、顶面、三角形梁架大部分均为原有木结构，施工工艺精细。

Railway Station Wooden Footbridge • Internal Structure

A square tall window at the upper part of facade provided good natural lighting inside. The wooden boards, side walls, top surfaces and triangle beams are all delicately executed.

【铁路俱乐部】

　　建于1906年，是修建中东铁路的俄国人娱乐、休闲的场所，现保存完整，建筑面积1736平方米，是昂昂溪建筑群中体积最大的。砖木结构，外观为俄式建筑风格，齐齐哈尔市一级保护建筑。

Railway Club

Built in 1906, it was an entertainment venue for Russian workers constructing the Middle East Railway. It is well preserved Russian style building of 1,736 square rneters and is the largest brick structure in Ang'angxi and is listed as a First Class Preserved Building of Qiqihar.

铁路俱乐部·立面构造与装饰

  俱乐部主体部分共2层，入口位于北面偏一隅。入口部分向外凸出，有弧形顶面；窗体均为竖向长窗，外部有环窗装饰；立面转角处装饰形态多样、层次丰富。

Railway Club • Facade and Decoration

The main part of the club is two storeys. The entrance, located at the north corner, extends outward and has a curved top. The façade is decorated with both tall, rectangular windows and ring windows. The corners of facade diversely and richly decorated.

> 昂昂溪俄式住宅

　　分布在昂昂溪道北的俄式建筑群中，住宅共有90多栋，均为田园式风格，拥有花园式的独户庭院。建筑有的是砖木结构，有的是用石头砌筑。

Russian Dwellings in Ang'angxi

There are over 90 Russian dwellings among the Russian building group located North of Ang'angxi Road. They are all single-family courtyards with gardens in the pastoral style. Some of them are wood and brick structures while others are made of stone.

昂昂溪俄式住宅·建筑特色

　　建筑多高举架、开间大跨度、人字屋架、大斜面铁皮屋顶，墙面上有用砖块砌成的凸凹花饰，丰富了立面和门窗造型。外墙面采用黄色粉刷，色彩明快，体现了俄罗斯的建筑风格。

Ang'angxi Russian Dwellings • Architectural Features

Most of the buildings have high, long span width eaves and beveled iron roofs. There are concave-convex floral ornaments of brick on the walls which enrich the designs of façade designs, windows and doors. The outer wall is painted in bright yellow in Russian style.

横道河子镇镇区

横道河子镇地处群山环绕之中。区内有佛手山、大石门、人头峰等国家级自然景观,又有中东铁路遗址、俄式教堂、木制俄式建筑等景观近200处,有"花园城镇"之称。

Downtown Hengdaohezi

Hengdaohezi is surrounded by mountains which include State-Level Natural Landscapes: Foshou Hill, Dashi Gate and Rentou Peak as well as another 200 landscapes including relics of the Middle East Railway, Russian church and Russian wooden structures. It is known as "Garden Town".

[横道河子火车站]

  始建于1901年，现为三等站。主站舍共2层，体态轻盈，体块明确，属典型俄式建筑风格，建筑最醒目的标志是两个巨大的红色铁质鱼鳞瓦锥形穹顶，是横道河子镇的代表性建筑。

Hengdaohezi Railway Station

It is a third-class station built in 1901. The main building has 2 storeys and is of typical Russian style. The most striking features of the building are two huge conical domes covered in red iron tiles. It is the representative architecture of Hengdaohezi.

圣母进堂教堂

始建于1901年，又叫约金斯克教堂，俗称"喇嘛台"，占地面积5000平方米，建筑面积400平方米。室内可以容纳500名信徒进行宗教活动，现国家级文物保护单位。

Church of the Presentation of the Blessed Virgin Mary

Built in 1901, it is also known as Vvedenskaya Church and as "Lama Tai". Its grounds cover an area of 5,000 square meters with a construction area of 400 square meters. The sanctuary can accommodate 500 worshippers. It is a State-Level Heritage Conservation Unit.

圣母进堂教堂·环境地势

建筑外部院落属回字形，入口处有三段式踏步，逐层高起的地势强调了建筑的气势。主立面正对入口，外部有一处镇区公众集散地。

Church of the Presentation of the Blessed Virgin Mary · Environment

The exterior courtyard of the building is behind the fountain and there are stairs to the entrance have three landings. The elevated terrain highlights the magnificence of the building. The main facade is opposite to the entrance and there is a public distribution center outside.

圣母进堂教堂·主厅立面

　　主立面纵向分为三段：下段是以竖向木板条为外部装饰；中段为原木木刻楞，每侧开三个大窗，窗额重点装饰；上段是从窗檐上部到屋檐下，墙体外部也用错拼的木板条装饰。

Church of the Presentation of the Blessed Virgin Mary · Main Facade
The main facade is vertically divided into three sections. The lower parts use vertical wooden slats for external decoration. The middle part is made of logs with three large heavily decorated windows on each side . The external walls are decorated with the staggered wooden slats.

[中东铁路机车库]

　　位于镇区西北,占地面积约5000平方米,机车库面积2160平方米。1903年中东铁路通车之后建成,是中东铁路在横道河子附近最大的一处机车库。国家级文物保护单位。

Engine Garage, Middle East Railway

Located in the northwest of the town, covering an area of about 5,000 square meters with a garage area of 2,160 square meters, it was built in 1903 after the Middle East Railway was opened to traffic and was the largest engine garage along the Middle East Railway near Hengdaohezi. It is a State-Level Heritage Conservation units.

中东铁路机车库·车库门

建筑体量庞大，为扇形全砖结构单层建筑，共有15个车库门并列组成，在正面看呈现出一波连浪的壮观景象，俄罗斯建筑风格明显。

Engine Garage, Middle East Railway • Garage Doors

This large, one-storey building made of fan-shaped bricks, has 15 garage doors standing side by side and it looks spectacular from the front view. It is of typical Russian style.

中东铁路机车库·破败的建筑与环境

虽已被列为保护建筑，但机车库已经遭受到了不同程度的破坏。扇形的向心方向是一片开阔地，由于时久不用已经杂草丛生。

Engine Garage, Middle East Railway • Dilapidated Building and Environment

Though listed as a Protected Building, the engine garage has suffered considerable damage. Abandoned for a long time, there is an overgrown field of weeds around the building

中东铁路机车库·区域入口道路

机车库区域内铺设有规则的，枕木形式的石质条板。此区域以机车库为主体，几类空间结构物质形式的组织与表现受控于建筑平面图形的构图作用，体现出场所存在的鲜明秩序。

Engine Garage, Middle East Railway · Road of Regional Entrance

The ground of the engine garage area is paved with regular stone slats.. Engine garage is the main building in this area but there are several types of structures of various materials forming distinctive site.

中东铁路大白楼

1904年为修建中东铁路的专家、技术人员建造的办公及住所，砖石结构，共3层，典型俄式建筑，因楼体外墙呈白色，当地俗称"大白楼"。国家级文物保护单位。

Dabailou (Grand White Building), Middle East Railway

Built in 1904, it was the office and residential building for the experts and technical staff working for the Middle East Railway. It was a 3-storey masonry structure typical Russian style. It was known locally as "Dabailou" (Grand White Building) because of its white walls. It is a State-Level Heritage Conservation Unit.

俄式木屋

　　位于镇北，现301国道西侧路基下面，为一排5栋俄罗斯式建筑的木房。木屋起初供中东铁路东段筑路指挥中心使用，后改为铁路职工住宅，在使用中得以保存。

Russian Wooden Houses
Located north of town, below the west roadbed of 301 State Road, are a row of five Russian-style, wooden houses initially used as the construction coordination center of the eastern section of the Middle East Railway. Later they became residences for the railway staff and continue as such today.

俄式木屋·木质结构和屋顶

木制房屋是俄罗斯传统建筑类型，多以板材配合厚重的混凝土与毛毡层，冬季温暖厚实，夏季凉爽舒畅。

Russian Wooden Houses· Wooden Structure and Roof

Wooden houses are traditional Russian buildings. Most are supplemented with concrete and felt layers making them warm in winter and cool in summer.

[民居院落·栅栏院]

　　小镇民居院落结合当地的实际情况形成了自身的特色：以栅栏院的形式使院落之间的关系更为密切，渗透性强，排布密实，多就近引水，按地形挖砌水渠。

Residential Courtyards • Fence Courtyard
The residential courtyards of the town have their own individual characteristics based on the local conditions. Lightly fenced courtyards create closer relationships between the courtyards. They have densely latticed fences. They have water diversion channels dug according to the terrain.

居院落·粗制的院门

　　一些住宅在较低的仓房旁边留出一个空隙，简单装上粗制的木板，成为住宅入口。木板门也只是象征性的关闭，没有设置锁具。院落位于此入口与建筑外围，直接毗邻镇区主干道。

Residential Courtyards • Rustic Courtyard Gates
Some residences simply installed crude wooden boards to form the outer entrances of the courtyards. Wooden doors were only symbolically closed instead of being locked. The courtyard is located outside of the entrance and the building, directly adjacent to the town's main roads.

【民居院落·讲究的外环境】

　　一些院落以低矮的水泥台作为边界，形成了虚拟的半封闭空间，各家院落彼此相邻，表现出一种经过人工修筑的秩序美。

Residential Courtyards · Exquisite External Environment

Some of the courtyards use low cement platforms as boundaries to form semi-enclosed space. All courtyards are adjacent to each other showing a kind of beauty of artificial construction.

后　　记

　　本书从实地调研、收集素材，到整理记录、精选入稿，到最终完成总共历时4年。介绍与表现黑龙江省特色地区建筑是我一直以来的心愿，回忆经历，感慨良多，而今得以实现，着实畅快淋漓。

　　由于特殊的地理位置，黑龙江省的建筑景观环境受到外来文化影响深远，且各地区呈现不同的表现效果：哈尔滨"洋"味十足，堪称世界建筑博览会；齐齐哈尔在自然中显露着腼腆，融自然环境、历史文化、人文景观为一体；横道河子镇饱经历史风霜，俄式建筑很大程度上得以保留了下来……在这片黑色的土地上，建筑景观犹如从历史的长河中绽放而来的七色花，充满着魅力。

　　在调研的过程中，一些具有特色的古建筑和景观没有得到很好地保护，已经有了相当程度的损毁。例如哈尔滨的圣伊维尔教堂，想来是十分痛心的。借此机会呼吁有关部门能够出台新的政策恢复古建筑原有的风采。另外，由于本书内容和版面的限制，部分特色建筑并没有列入其中，之前考察期间完成的画作也有很多没有机会展示出来，只能算作一件憾事了。

　　本书的具体描述与画作的表现亦有不足之处，恳请广大读者和同行予以指正，在此深表谢意。

王安安
2013年3月